Sami El Khatib
Maha El Khatib

Visual Perception, Signal Processing, & Associated Eye Diseases

Sami El Khatib
Maha El Khatib

Visual Perception, Signal Processing, & Associated Eye Diseases

Emphasize on Human Eye & Its Diseases

LAP LAMBERT Academic Publishing

Imprint
Any brand names and product names mentioned in this book are subject to trademark, brand or patent protection and are trademarks or registered trademarks of their respective holders. The use of brand names, product names, common names, trade names, product descriptions etc. even without a particular marking in this work is in no way to be construed to mean that such names may be regarded as unrestricted in respect of trademark and brand protection legislation and could thus be used by anyone.

Cover image: www.ingimage.com

Publisher:
LAP LAMBERT Academic Publishing
is a trademark of
International Book Market Service Ltd., member of OmniScriptum Publishing Group
17 Meldrum Street, Beau Bassin 71504, Mauritius

Printed at: see last page
ISBN: 978-620-0-30692-0

Visual Perception, Signal Processing, & Associated Eye Diseases

Emphasize on Human Eye & Its Diseases

Sami El Khatib

Vice Dean – School of Arts & Sciences
Ph.D Cell Biology & Oncology
Lebanese International University
Department of Biological Sciences
Bekaa Campus
Khiyara – West Bekaa
Lebanon
Tel   +961 76012570
E-mail : sami.khatib@liu.edu.lb

Maha El Khatib

Msc. Applied Biotechnology
Lebanese University
Department of Biological Sciences
Faculty of Sciences
Lebanon
E-mail : maha.ak94@hotmail.com

Contents

I. Introduction

Humans possess the remarkable ability to perceive color, shape and motion, and to differentiate between light intensities varied by over nine orders of magnitude. This ability is mediated by the marvelous organ of vision, the eye.

The eye is an optical device that transmits and focuses light into the retina. The retina, ~ 0.2 mm thick central nervous tissue, is the first station of the visual system. In addition to acting as a light receiver, the retina carries out considerable image processing through circuits that involve five main classes of cells (i.e., photoreceptors, bipolar cells, amacrine cells, horizontal cells, and ganglion cells). These Processes collectively amplify, extract, and compress signals to preserve relevant information before it gets transmitted for further processing in the central nervous system (the brain) through the optical nerves. The retinal information received by the brain is processed to control eye movement, pupil size, and circadian rhythm.

The human visual system is so extraordinary in the quantity and quality of information it supplies about the world. A glance is sufficient to describe the location, size, shape, color, and texture of objects and, if the objects are moving, their direction and speed. Equally remarkable is the fact that visual information can be discerned over a wide range of stimulus intensities, from the faint light of stars at night to bright sunlight.

This book describes the process of vision and visual processing, and it ends up with describing some of the most common eye diseases and disorders as well as highlighting on some food kinds that enhance vision.

II. The Human Eye: Anatomy and Terminology

A sensory system is a part of the nervous system that is responsible for processing sensory information. A sensory system consists of sensory receptors, neural pathways, and parts of the brain involved in sensory perception. Commonly recognized sensory systems are those for vision, hearing, somatic sensation (touch), taste, and olfaction (smell). Among all these, vision is the most fundamental of our senses.

Our eyes are marvelous sense organs that make it possible for us to see. They allow us to appreciate all the beauty of the world we live in, to read and gain knowledge, and to communicate our thoughts and feelings to each other through visual expression and visual arts.

II.1. Anatomy of the Human Eye

When looking into someone's eyes, we may identify the following different structures (**Figure 1**):

- o **A black-looking aperture**, the **pupil**, which is located at the center of the eye and allows light to enter the eye. It appears dark because of the absorbing pigments in the retina (Kolb, 2013).
- o **A colored circular muscle**, the **iris**, which is beautifully, pigmented giving us our eye's color (the central aperture of the iris is the pupil). This circular

muscle controls the size of the pupil so that more or less light, depending on conditions is allowed to enter the eye. Eye color, or more correctly, iris color is due to variable amounts of eumelanin (brown/black melanins) and pheomelanin (red/yellow melanins) produced by melanocytes. The majority of people are brown eyed, and the rest are blue and green-eyed (Kolb, 2013).

- o **A transparent external surface**, the **cornea**, which covers both the pupil and the iris. This is the first and the most powerful lens of the optical system. It allows, together with the crystalline lens, the production of a sharp image at the retinal photoreceptor level (Widmaier, 2008).

- o **The "white of the eye"**, the **sclera**, which forms part of the supporting wall of the eyeball. It forms a white capsule around the eye except at its anterior surface where it is specialized into the clear cornea. The sclera is continuous with the cornea. Furthermore, this external covering of the eye is in continuity with the dura matter of the central nervous system (Widmaier, 2008).

This tough, fibrous sclera serves as the insertion point for external muscles that move the eyeball within their sockets.

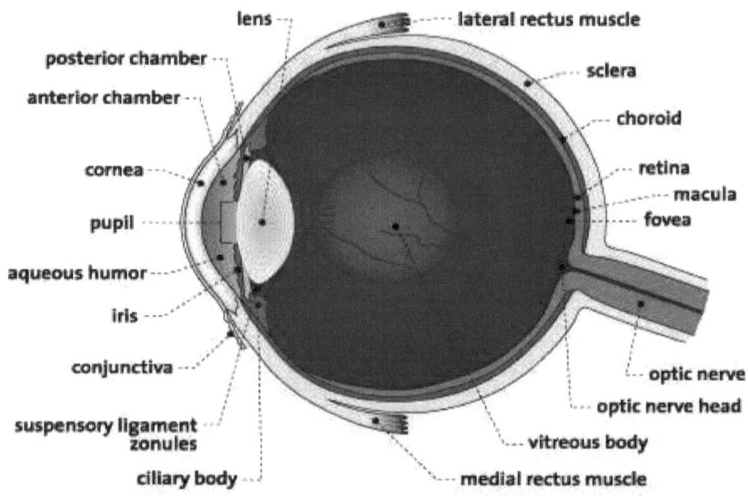

Figure 1: The human eye, cross sectional view showing internal structure

Underlying the sclera is the **choroid**, a darkly pigmented layer that absorbs light rays at the back of the eyeball (Kolb, 2013).

Deeper inside the eye is the lens, which further refracts light and helps create a finer image. The **crystalline lens** is made of layers of a fibrous material and has an index of refraction of roughly 1.40. Unlike the lens of a camera, the lens of the eye is able to change its shape and thus serves to adjust the visual process. The lens is attached to the **ciliary muscles** and suspended by ligaments (called **zonule fibers**) attached to the anterior portion of the ciliary muscles. The contraction or relaxation of these ligaments as a consequence of ciliary muscle actions changes the shape of the lens. This process is called **accommodation.** It allows us to form a sharp image on the retina (Kolb, 2013).

The inner surface of the eye is known as the **retina**. It is an extension of the brain that lines the inner-posterior surface of the eye and contains numerous types of neurons as well as the eye's sensory cells, called photoreceptors. These photoreceptor cells are the rods and cones. An adult eye is typically equipped with up to 120 million rods that detect the intensity of light and about 6 million cones that detect the frequency of light. These rods and cones send nerve impulses to the brain. The nerve impulses travel through a network of nerve cells. There are as many as one million neural pathways from the rods and cones to the brain. This network of nerve cells is bundled together to form the **optic nerve** on the very back of the eyeball (Widmaier, 2008).

The point at the back of the eye at which optic nerves exit the eye is called the **blind spot**. This region of the retina contains no photoreceptors and thus, it is insensitive to light. Objects that lie completely within the blind spot are not perceived (Kolb, 2013).

The central point for image focus in the human retina is the **fovea**. It is a depression in the retina that contains only cones and that provides us with acute eye sight (Kolb, 2013).

II.1.a. Three Different Eye Layers

A cross-sectional view of the eye shows that it contains three different layers (Kolb, 2013):

- The external layer, formed by the sclera and cornea
- The intermediate layer, and is divided into two parts: the anterior (iris and ciliary body) and the posterior (choroid)
- The internal layer, or the sensory part of the eye, the retina

II.1.b. Chambers Found in the Human Eye

The eye also contains three chambers of fluid (Widmaier, 2008):

- The anterior chamber (between the cornea and the iris),
- The posterior chamber (between the iris, suspensory ligament zonules, and the lens)
- The vitreous chamber (between the lens and the retina).

The first two chambers are filled with aqueous humor whereas the vitreous chamber is filled with a more viscous fluid, the vitreous humor.

II.1.c. Supporting Parts of the Human Eye

Other parts of the human eye play a supporting role in the main activity of sight. Some carry fluids (such as tears and blood) to lubricate or nourish the eye, others are muscles that allow the eye to move, some parts protect the eye from injury (such as the lids and the epithelium of the cornea), and some are messengers, sending sensory information to the brain (such as the pain-sensing nerves in the cornea and the optic nerve behind the retina).

Each part of the eye plays a distinct role in enabling humans to see. The ultimate goal of such an anatomy is to allow humans to focus images on the back of the eye, the retina. Although all parts of the eye are important for perceiving a good image, the most vital layer for vision is the retina (Widmaier, 2008).

II.2. Light and vision

Our eyes are sensitive only to a tiny portion of the vast spectrum of electromagnetic radiations, that we call visible light. Radiant energy is expressed in terms of wavelength (λ) and frequency. Those wavelengths capable of stimulating the photoreceptors of the eye range from a bit less than 400 nm to about 750 nm. Different wavelengths within the visible spectrum are perceived as different colors (**Figure 2**) (Widmaier, 2008).

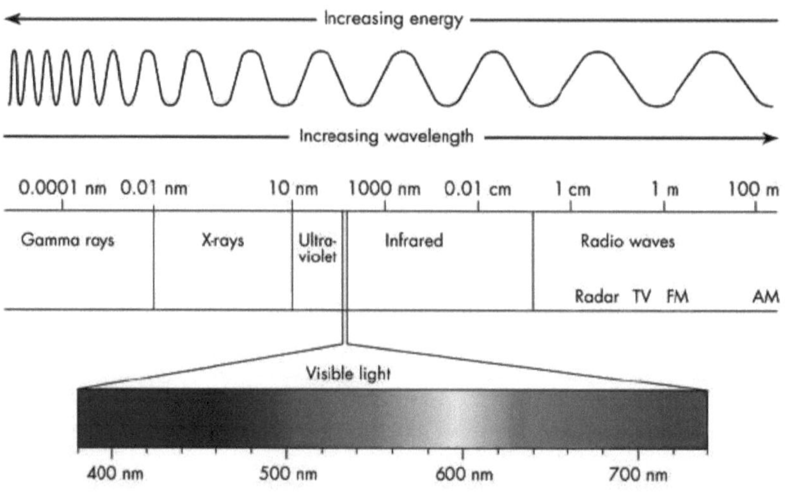

Figure 2: The electromagnetic spectrum. Visible light ranges from 400 to 750 nm

(http://lumenistics.com/what-is-full-spectrum-lighting/).

III. The Retina

Vision begins with the capture of light energy by photoreceptors present in the human retina. The retina is about 0.5 mm thick. It is the neural portion that lines the inner posterior part of the human eye (Kolb, 2013). It is responsible for the conversion of light signals, provided by the incoming photons, into electrical signals, which are sent to the brain via the optic nerve for procession and thus, for image perception. Therefore, it is not surprising to study the anatomy and physiology of the retina before understanding the process of visual perception (Purves, 2004).

III.1. Embryonic Retinal Development

The retina is part of the central nervous system. It is derived from the neural tube and is formed during development of the embryo from optic vesicles out pouching from two sides of the developing neural tube. The primordial optic vesicles fold back in upon themselves (invaginate) to form the optic cup with the inside of the cup becoming the retina and the outside remaining a single monolayer of epithelium known as the retinal pigment epithelium (RPE). Initially both walls of the optic cup are one cell thick, but the cells of the inner wall divide to form a neuro-epithelial layer many cells thick: the retina (**Figure 3**) (Kolb, 2013).

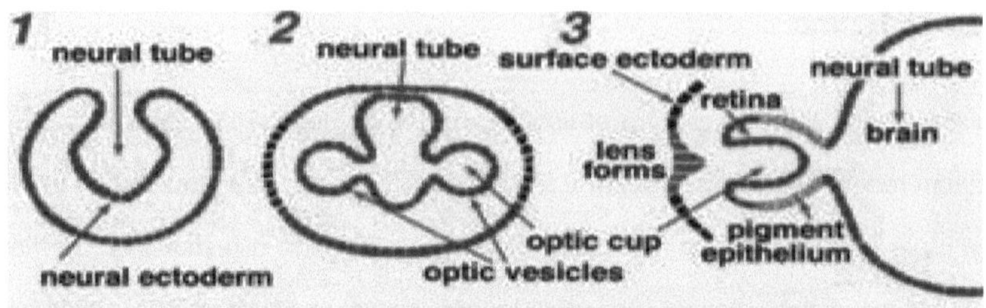

Figure 3: Development of the human eye. The retina is derived from the neural tube in which the optic vesicles invaginate forming the optic cup. The inside of the cup becomes the retina while the outside becomes the retinal pigment epithelium (Kolb, 2013).

So, retinal development is characterized by the formation of many retinal layers arising from cell division and subsequent cell migration. The retina develops in an inside to outside manner (**Figure 4**), that is, ganglion cells are formed first and photoreceptor cells become fully mature last (Purves, 2004).

III.2. Cell Population in the Retina

The human retina consists of neurons of more than 60 different types which can be grouped into five distinct families of neurons: photoreceptors (rods and cones), bipolar cells, horizontal cells, amacrine cells and ganglion cells (Masland, 2012). The cell bodies and processes of these neurons are stacked in alternating layers, with the cell bodies located in the inner nuclear, outer nuclear, and ganglion cell layers, and the processes and synaptic contacts located in the inner plexiform and outer plexiform layers (**Figure 4**) (Purves, 2004).

The major route of information flow through the human retina is from photoreceptor cells to bipolar cells to ganglion cells; a three neuron chain. The neurons in the visual flow route are often describes as first order neurons (photoreceptors), second order neurons (horizontal cells and bipolar cells), and third order neurons (amacrine cells and ganglion cells).

III.2.a. Photoreceptors

There are two types of photoreceptors in the retina: rods and cones. You might ask about the reason behind having two types of photoreceptors. The answer is that the rods are absent from the fovea and that night vision is poor when incoming light is confined to this region. So, the rods are responsible for night vision, while the cones are responsible for day/color vision (Schiller, 2010).

Both rods and cones have an outer segment composed of membranous disks that contain light-sensitive photopigment and lie adjacent to the pigment epithelium, and an inner segment that contains the cell nucleus and gives rise to synaptic terminals that contact bipolar or horizontal cells (**Figure 5**).

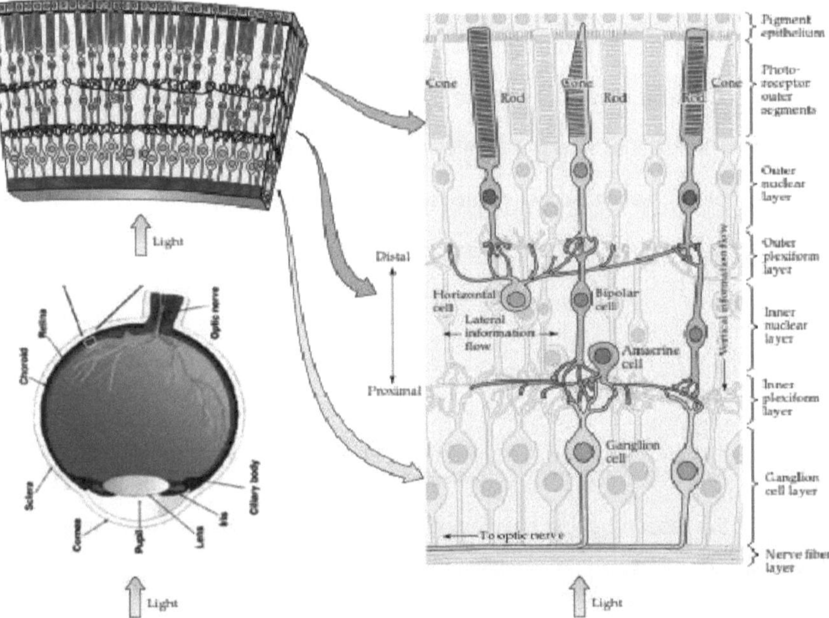

Figure 4: Structure of the retina. (A) Section of the retina showing overall arrangement of retinal layers. (B) Diagram of the basic circuitry of the retina.

A three-neuron chain -photoreceptor, bipolar cell, and ganglion cell — provide the most direct route for transmitting visual information to the brain. Horizontal cells and amacrine cells mediate lateral interactions in the outer and inner plexiform layers, respectively. The terms inner and outer designate relative distances from the center of the eye (inner, near the center of the eye; outer, away from the center, or toward the pigment epithelium) (Purves, 2004).

Absorption of light by the photopigment in the outer segment of the photoreceptors initiates a cascade of events that changes the membrane potential of the receptor, and therefore the amount of neurotransmitter released by the photoreceptor synapses onto the cells they contact. The synapses between photoreceptor terminals and bipolar cells (and horizontal cells) occur in the outer plexiform layer; more specifically, the cell bodies of photoreceptors make up the outer nuclear layer, whereas the cell bodies of bipolar cells lie in the inner nuclear layer. The short axonal processes of bipolar cells make synaptic contacts in turn on the dendritic processes of ganglion cells in the inner plexiform layer. The much larger axons of the ganglion cells form the optic nerve and carry information about retinal stimulation to the rest of the central nervous system (Purves, 2004).

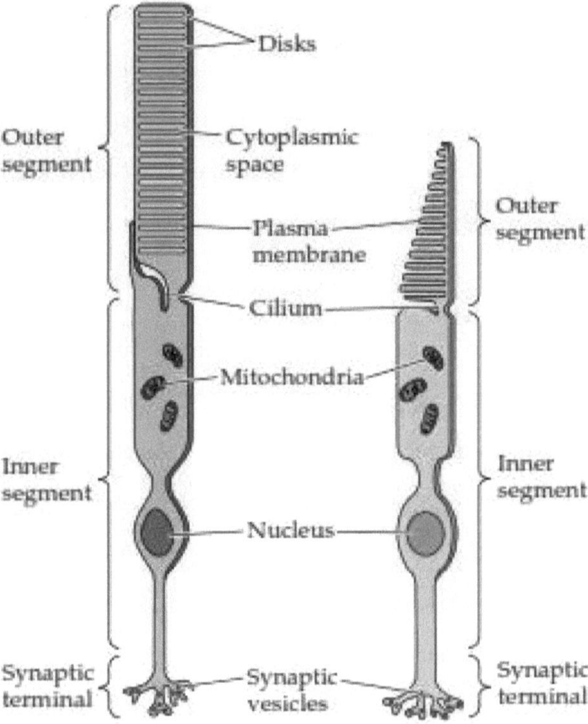

Figure 5: The structure of a photoreceptor. A photoreceptor is made up of an outer segment (close to the retinal pigment epithelium), an inner segment, and a synaptic terminal. A rod is shown on the left and a cone on the right (Purves, 2004).

III.2.b. The Inter-Neurons: Bipolar, Horizontal, & Amacrine Cells

The inter-neurons in the retina, bipolar cells, horizontal cells, and amacrine cells, serve to connect the photoreceptors with the ganglion cells. They have their cell bodies in the inner nuclear layer and have processes that are limited to the outer (horizontal cells), inner (amacrine cells), and both outer and inner plexiform layers (bipolar cells) (**Figure 4**) (Hildebrand, 2011).

The processes of horizontal cells enable lateral interactions between photoreceptors and bipolar cells that maintain the visual system's sensitivity for light intensities. They provide inhibitory feedback to rods and cones and possibly to dendrites of bipolar cells through secreting the inhibitory neurotransmitter γ-amino butyric acid (GABA) (Misland, 2012).

The processes of amacrine cells are postsynaptic to bipolar cell terminals and presynaptic to the dendrites of ganglion cells. Different subclasses of amacrine cells are thought to make distinct contributions to visual function (Purves, 2004). The most important amacrine cells are the starburst amacrine cells that are involved in motion detection (Hildebrand, 2011) and the AII amacrine cells that are involved in scotopic (nocturnal) vision. The variety of amacrine cell subtypes illustrates the more general rule that although there are only five basic retinal cell types, a considerable diversity can be observed within a given cell type (Purves, 2004).

In the simplest case, the photoreceptor cell is directly connected to a ganglion cell via a bipolar cell. Bipolar cells receive input from either rods or cones. Cone bipolar cells may make contact with as few as one cone, while rod bipolar cells may receive input from up to 70 rods (Hildebrand, 2011).

Bipolar cells have concentric receptive fields and are classified as ON- and OFF-type bipolar cells from the response of the receptive field center (Kaneda, 2013). ON-type bipolar cells depolarize when their receptive field center is illuminated and hyperpolarize when their receptive field surround is illuminated. OFF-type bipolar cells hyperpolarize when their receptive field center is illuminated and depolarize when their receptive field surround is illuminated. ON-type bipolar cells include 5 subtypes of ON-cone bipolar cells and rod bipolar cells, whereas OFF-type bipolar cells have 4 subtypes of OFF-cone bipolar cells only (**Figure 6**) (Kaneda, 2013).

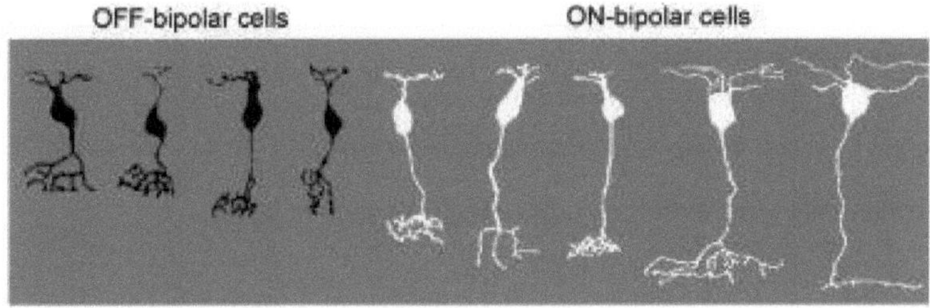

Figure 6: The ON- and OFF- bipolar cells. There are 4 OFF-cone bipolar cells and 5 ON-bipolar cells (for rods and cones) (Sterling, 2013).

The difference between ON and OFF responses is due to the expression of two classes of glutamate receptor. OFF bipolar cells express AMPA and kainate type receptors, which are cation channels opened by glutamate; since photoreceptor cells hyperpolarize in response to light, these bipolar cells hyperpolarize in response to light as well, because less glutamate arrives from the cone synapse. ON bipolar cells express mGluR6, a metabotropic receptor, which, when glutamate binds to the receptor, leads to closing of the cation channel TRPM1. The receptor is thus sign inverting. When light causes less glutamate to be received from the photoreceptor terminal, cation channels open and the cell depolarizes (Mesland, 2013).

The center response is mediated by glutamate released from photoreceptors, whereas the surround response is thought to involve GABAergic feedback to bipolar cells mediated by horizontal cells (Kaneda, 2013).

Thus, on their way to the ganglion cell, visual signals are transmitted and modified by bipolar, horizontal, and amacrine cells as part of the visual processing within the retina.

Ganglion cells are those cells responsible of transmitting visual information from the retina to the brain for further processing (Hildebrand, 2011).

The ganglion cell bodies are located in the ganglion cell layer, and their dendrites make contact with bipolar and amacrine cells in the inner plexiform layer (IPL). Up to 20 different ganglion cell types have been described in the human retina. The two best known types are the midget and the parasol cells, which make up about 80% of ganglion cell population. The midget ganglion cell (also known as P or B cell) is a small cell with a relatively small dendritic arbor. The Parasol ganglion cell (M or A cell) has a much more extensive dendritic arbor that resembles an opened umbrella in histological preparations of the retina. Midget and parasol cells project to the parvocellular and magnocellular layers of the lateral geniculate nucleus (LGN) in the thalamus, respectively (Hildebrand, 2011).

III.3. Central and Peripheral Retina

Central retina is that portion of the retina close to the fovea. It is much thicker than peripheral retina (at the periphery) due to increased packing of photoreceptors, particularly cones, along with the associated inter-neurons and ganglion cells (Kolb, 2013).

Central retina is cone-dominated retina whereas peripheral retina is rod-dominated. Thus, the inner nuclear layer (INL) is thicker in the central retina compared with peripheral retina, due to a greater density of cone-connecting second-order neurons (cone bipolar cells and horizontal cells) and third order amacrine cells concerned with the cone pathways (Kolb, 2013).

IV. Visual Perception

As we have previously mentioned, vision begins with the absorption of light by visual pigments in the retinal rod and cone photoreceptors. This is followed by the conversion of light energy into electric energy, a process called photo-transduction.

The two kinds of photoreceptors, rods and cones, are so named for their overall shapes (Luo, 2008). Cones are robust conical shaped structures, whereas rods are slim, rod shaped structures (**Figure 7**) (Kolb, 2013).

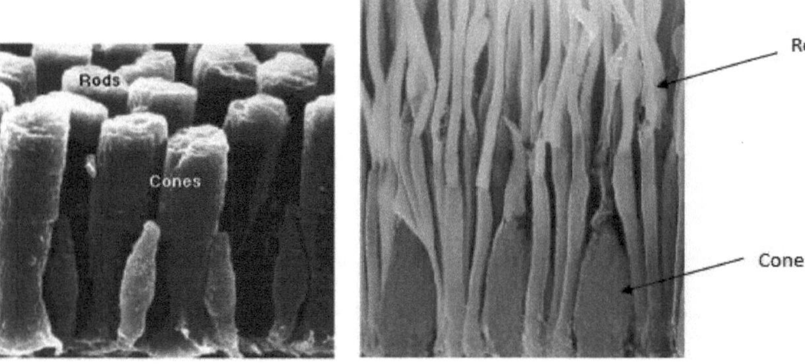

Figure 7: Scanning electron micrographs of the rods and cones of the human retina (Kolb, 2013).

Rods and cones are specialized unipolar neurons. All vertebrate visual receptors follow a simple blueprint. They can be divided into two portions, termed inner and outer segments, according to their radial position within the retina. The inner segment consists of the cell body and contains the cellular organelles found in other neurons, including a synaptic terminal. The outer segment is an elaborate, modified cilium that contains the biochemical machinery needed for visual transduction. The components of the phototransduction enzyme cascade are packed into stacks of flattened, membranous vesicles ("disks") (**Figure 8**) that in rods are enclosed by the plasma membrane of the outer segment. Cones have a similar structure, but the disk membranes are an extension of the plasma membrane, arranged into a series of enfolding (Isayma, 2013).

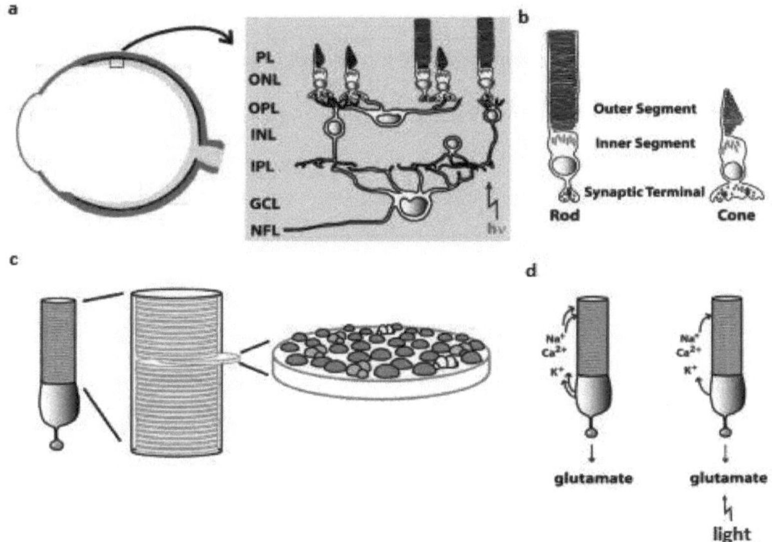

Figure 8: Cellular plan of vertebrate photoreceptors. **a.** the neural retina within the vertebrate eye. **b.** Morphology of rods and cones. **c.** Outer segment disk membranes containing the biochemical components of the light-detecting pathway; photo pigments. **d.** The "dark" current. Na$^+$ and a lesser amount of Ca^{2+} enter through cyclic nucleotide-gated channels in the outer segment membrane while K$^+$ departs through voltage-gated channels in the inner segment. In darkness (left), the rod is depolarized and releases the neurotransmitter glutamate continuously. In response to light (right), channels in the outer segment membrane close, the rod hyperpolarizes, and glutamate release decreases (Isayama, 2013).

IV.1. Anatomical Distribution of Rods and Cones

The distribution of rods and cones across the surface of the retina has important consequences for vision (**Figure 9**). Despite the fact that perception in typical daytime light levels is dominated by cone-mediated vision, the total number of rods in the human retina (about 90 million) far exceeds the number of cones (roughly 4.5 million). As a result, the density of rods is much greater than cones throughout most of the retina. However, this relationship changes dramatically in the **fovea**, a highly specialized region of the central retina that measure about 1.2 millimeters in diameter (Purves, 2004).

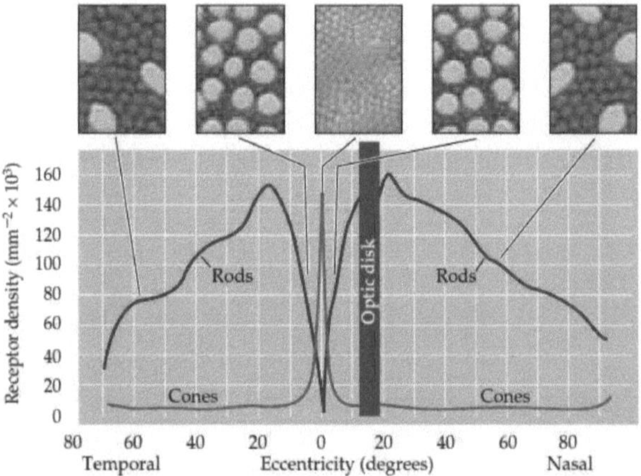

Figure 9: Distribution of rods and cones in the human retina. Graph illustrates that cones are present at a low density throughout the retina, with a sharp peak in the center of the fovea. Conversely, rods are present at high density throughout most of the retina, with a sharp decline in the fovea. Boxes at top illustrate the appearance of face on sections through the outer segments of the photoreceptors at different eccentricities. The increased density of cones in the fovea is accompanied by a striking reduction in the diameter of their outer segments (Purves, 2004).

In the fovea, cone density increases almost 200-fold, reaching, at its center, the highest receptor packing density anywhere in the retina. This high density is achieved by decreasing the diameter of the cone outer segments such that foveal cones resemble rods in their appearance. The increased density of cones in the fovea is accompanied by a sharp decline in the density of rods. In fact, the central 300 µm of the fovea, called the **foveola**, is totally rod-free. The extremely high density of cone receptors in the fovea, and the one-to-one relationship with bipolar cells and retinal ganglion cells, awards this component of the cone system with the capacity to mediate high visual acuity. As cone density declines with eccentricity and the degree of convergence onto retinal ganglion cells increases, acuity is markedly reduced. Just 6° eccentric to the line of sight, acuity is reduced by 75%. The restriction of highest acuity vision to such a small region of the retina is the main reason humans spend so much time moving their eyes (and heads) around, in effect directing the foveas of the two eyes to objects of interest (Purves, 2004).

Another anatomical feature of the fovea (which literally means "pit") that contributes to the superior acuity of the cone system is that the layers of cell bodies and processes that overlie the photoreceptors in other areas of the retina are displaced around the fovea, and especially the foveola (**Figure 10**). As a result, photons are subjected to a minimum of scattering before they strike the photoreceptors. Finally, another potential source of optical distortion that lies in the light path to the receptors—the retinal blood vessels—are diverted away from the foveola. This central region of the fovea is therefore dependent on the underlying choroid and pigment epithelium for oxygenation and metabolic sustenance (Purves, 2004).

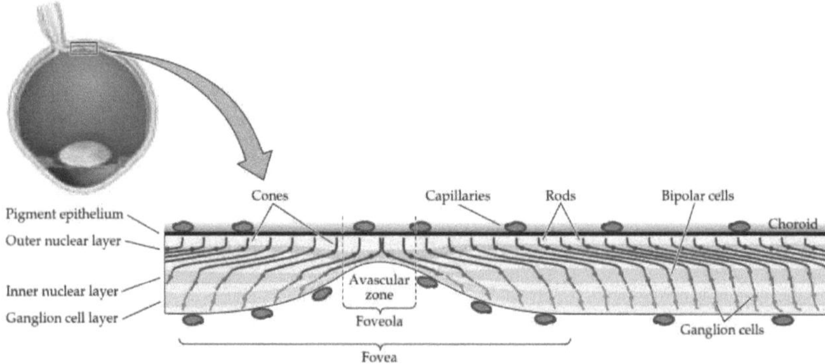

Figure 10: Diagrammatic cross section through the human fovea. The overlying cellular layers and blood vessels are displaced so that light is subjected to a minimum of scattering before photons strike the outer segments of the cones in the center of the fovea, called the foveola (Purves, 2004).

IV.2. Cones and Color Vision

A special property of the cone system is color vision. Perceiving color allows us to discriminate objects on the basis of the distribution of the wavelengths of light that they reflect to the eye. Color obviously gives us a quite different way of perceiving and describing the world we live in.

Unlike rods, which contain a single photopigment, there are three types of cones that differ in the photopigment they contain. Each of these photopigments has a different sensitivity to light of different wavelengths, and for this reason are referred to as "blue," "green," and "red" or, more appropriately, short (S ~430 nm), medium (M ~530 nm), and long (L ~560 nm) wavelength cones—terms that more or less describe their spectral sensitivities (**Figure 11**). This nomenclature implies that individual cones provide color information for the wavelength of light that excites them best. In fact, individual cones, like rods, are entirely color blind in that their response is simply a reflection of the number of photons they capture, regardless of the wavelength of the photon (or, more properly, its vibration energy) (Purves, 2004).

Much of the information about color vision has come from studies of individuals with abnormal color detecting abilities. Color vision deficiencies result either from the inherited failure to make one or more of the cone pigments or from an

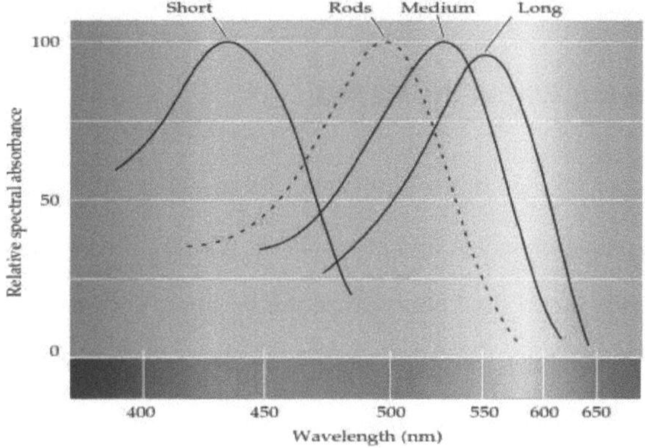

Figure 11: Color vision. The light absorption spectra of the four photopigments in the normal human retina. (Recall that light is defined as electromagnetic radiation having wavelengths between ~400 and 700 nm.) The solid curves indicate the three kinds of cone opsins; the dashed curve shows rod rhodopsin for comparison (Purves, 2004).

alteration in the absorption spectra of cone pigments. Under normal conditions, most people can match any color in a test stimulus by adjusting the intensity of three superimposed light sources generating long, medium, and short wavelengths. The fact that only three such sources are needed to match (nearly) all the perceived colors is strong confirmation of the fact that color sensation is based on the relative levels of activity in three sets of cones with different absorption spectra. That color vision is **trichromatic** was first recognized by Thomas Young at the beginning of the nineteenth century (thus, people with normal color vision are called trichromats) (Purves, 2004).

IV.3. Visual Processing

IV.3.a. Visual Processing Within the Retina

Most neurons maintain a resting membrane potential of about -60 to -70 mV and when excited, they open cation channels and allow Na^+ to flow in. The resulting depolarization opens voltage-gated Ca^{2+} channels at the synapse. Ca^{2+} flows in and promotes fusion of synaptic vesicles which release neurotransmitter (Isayama, 2013).

Rods and cones work "backwards". At rest, that is in darkness, rods and cones are depolarized to -35 to -45 mV (**Figure 12**) because channels in the outer segment membrane that pass Na^+ open. An efflux of K^+ through voltage-gated channels at the inner segment balances the influx of cations at the outer segment, completing an electrical circuit, referred to as the "dark" or circulating current. Researchers have discovered that cyclic guanosine monophosphate (cGMP) is necessary to keep the Na$^+$ channels open, and that the channels will close if the cGMP is converted into GMP (Isayama).

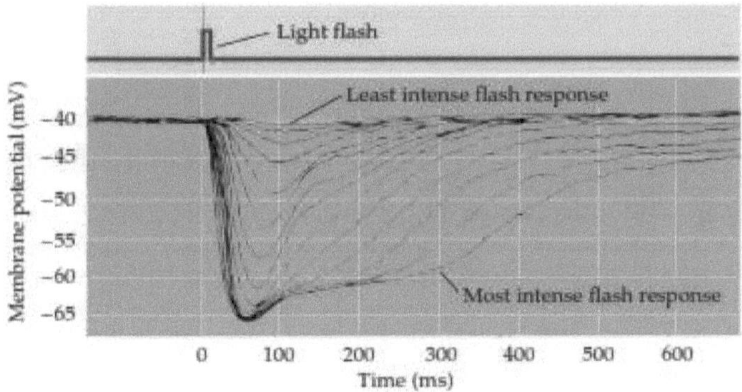

Figure 12: An intracellular recording from a single cone stimulated with different amounts of light. Each trace represents the response to a brief flash that was varied in intensity (Purves, 2004)

Since photoreceptors are already depolarized in darkness, Ca^{2+} channels are open and the influx of Ca^{2+} ions supports a steady release of the neurotransmitter glutamate from rod and cone terminals onto second order neurons. When rods and cones receive light, the channels in their outer segments close and the membrane hyperpolarizes towards the equilibrium potential for K$^+$. At the synapse, voltage-sensitive Ca^{2+} channels close and neurotransmitter release subsides (Schwartz, 2002).

The visual pigment is a G-protein coupled receptor (GPCR) that consists of a protein, opsin, covalently attached to a vitamin A-derived chromophore, 11-cis retinal (Kefalov, 2012). Retinal cannot be made *de novo* by the body and must be taken in constantly in the diet. Consequently, vitamin A deficiency causes night blindness; depletion of 11-*cis* retinal leads to fewer functional rhodopsins and lower sensitivity of the rods (Isayama, 2013).

Like all GPCRs, rhodopsin, the rod photopigment, consists of seven transmembrane segments (from H1 to H7) connected by three extracellular and three cytoplasmic loops (**Figure 13**) (Shichida, 2003), and the corresponding G protein is called transducin (Isayama, 2013).

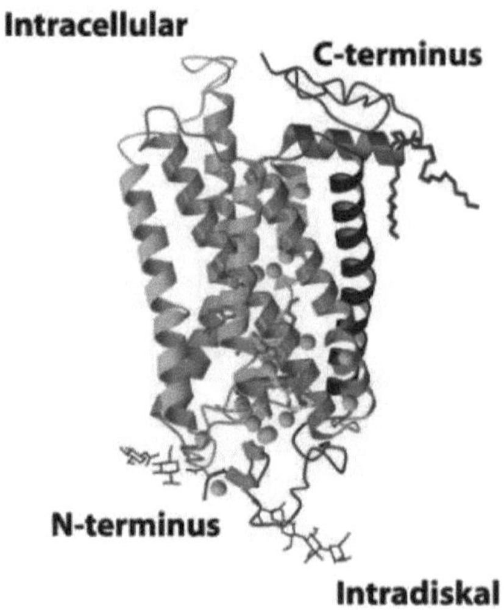

Figure 13: Structure of rhodopsin. X-ray crystal structure of rhodopsin (Isayama, 2013).

Transducin is heteromeric, consisting of α, β and γ subunits (**Figure 14**). In its inactive state, transducin's α subunit has a GDP bound to it, but active rhodopsin binds transducin and catalayzes the exchange of a bound GDP for a GTP, leading to the release of the α-GTPsubunit from transducin. That is because GDP acts like glue that holds the transducin heterotrimer together, so without GDP, the subunits come off (Isayama, 2013).

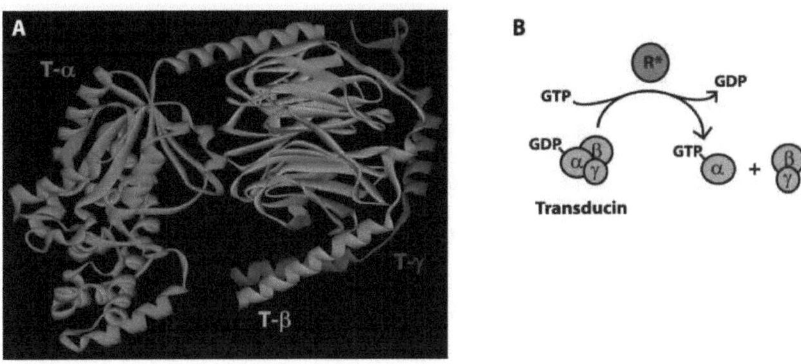

Figure 14: Structure of transducin. **A.** Heterotrimeric composition of transducin **B.** Activation of transducin. Active rhodopsin (R*) binds the transducin heterotrimer and catalyzes the exchange of a bound GDP for a GTP, leading to the release of the α-GTP subunit from transducin (Isayama, 2013).

The chromophore 11-cis retinal is covalently linked through a schiff base to a conserved lysine residue (Lys 296 acts as a linkage site for the chromophore and is contained within H7 in all pigments) (Shichida, 2003). A carboxylic acid residue is conserved within H3 and serves as the counter ion to the protonated, positively charged Schiff base (Menon, 2001).

The structure of 11-*cis* retinal makes it well suited to serve as the light-absorbing molecule in rod and cone pigments. Whenever a string of carbons are connected with alternating double and single bonds, the bonds take on a special character, not exactly a single bond but not exactly a double bond either. Furthermore, the bonding electrons do not stay with a particular carbon pair, rather they delocalize over the entire conjugated chain. It does not take too much energy to raise the electrons to the excited state. By itself, 11-*cis* retinal would absorb near UV light. But opsin perturbs the distribution of the electrons so that the excited state can be achieved with less energy, i.e., with longer wavelength light. Cone opsins have similar hepta-helical structures, but with different amino acid residues surrounding the bound 11-*cis* retinal; thus they tune the chromophore's absorption to different wavelengths (Isayama, 2013).

The human retina has one type of rod for dim light vision and three types of cone cells that allow color discrimination (Kefalov, 2012). Despite similarity in function, rods and cones exhibit important functional differences that can be demonstrated physiologically. First, rods are so sensitive that they can detect a single photon of light, which makes them perfectly suited for dim light vision. On the other hand, cones are up to 100-fold less sensitive than rods. As a result, they cannot signal in dim light conditions, depriving us of color vision at night. Second, rods saturate in even moderately bright light and remain nonfunctional during most of the day. In contrast, cones have a remarkable ability to adjust their sensitivity and remain photosensitive even in extremely bright light. This process, known as light adaptation, prevents cones from saturating in bright light and allows us to see throughout the day. With rods saturated, cones are responsible for most of the visual information reaching our brain during the day. Third, rods experience a long refractory time following exposure to bright light and can take up to 1h to completely recover their sensitivity. In contrast, cones quickly recover their sensitivity within a few minutes. This process, known as dark adaptation, prevents cones from becoming refractory and allows us to retain visual perception in a quickly changing light environment (Kefalov, 2012).

Upon absorbing a photon, 11-cis-retinal isomerizes to all-trans-retinal (**Figure 15**) (this is the only light dependent reaction in vision). As a result of this conformational change, retinal no longer fits in the binding site in opsin, and opsin then undergoes a series of spontaneous conformational changes to become active (in its Meta II state) (Luo, 2008).

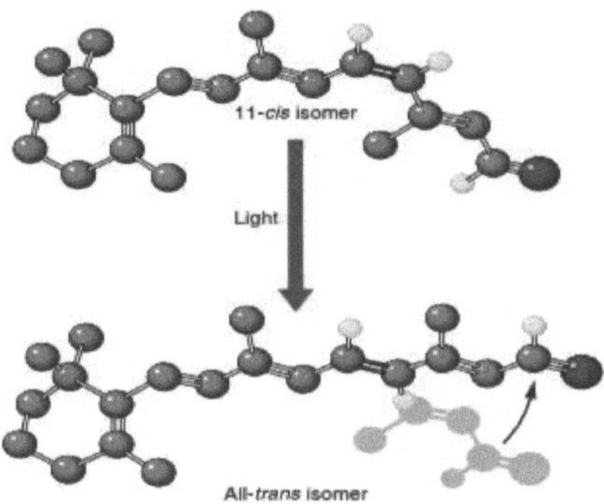

Figure 15: Absorption of light by retinal. When light is absorbed by a photopigment, the 11-cis isomer of retinal undergoes a change in shape forming 11- trans retinal. This change in retinal's shape initiates a chain of events that leads to hyperpolarization of the photoreceptor.

The primary opsin intermediate formed is photorhodopsin. Photorhodopsin is then thermally converted into many intermediates including Bathorhodopsin (Batho), blue-shifted intermediate (BSI), lumirhodopsin (Lumi), and metarhodopsin (I, II, and III), each of which exhibits a characteristic absorption spectrum (Shichida, 2003). Among all these intermediates, Meta II is the only active one, i.e, it is the only intermediate capable of activating transducin through catalyzing the GDP/GTP exchange reaction. Eventually the pigment is hydrolyzed into opsin and all-*trans*-retinal, and is said to be "bleached". Functional pigment is regenerated when opsin recombines with another 11-*cis*-retinal molecule (Luo, 2008).

Transducin serves as an intermediary between the activated receptor and an effector. In this case, the effector is a phosphodiesterase (PDE) that enzymatically cleaves the phosphodiester bond of cGMP to form 5'-GMP. PDE is a heterotetramer that consists of a dimer of two catalytic subunits, α and β, each with an active site inhibited by a PDE γ subunit. The activated transducin subunit α-GTP binds to PDE γ and relieves the inhibition on a catalytic subunit. All of these events take place within the disk membrane, separate from the ion channels in the plasma membrane of the outer segment. A small, soluble molecule, cGMP, links the two compartments. When activated, PDE hydrolyzes cGMP to 5'-GMP, the cGMP concentration inside the rod decreases, and cyclic nucleotide-gated ion channels respond by closing (**Figure 16 & 17**). Na^+ entry into the outer segment is blocked, thus the rod hyperpolarizes (Isayama, 2013).

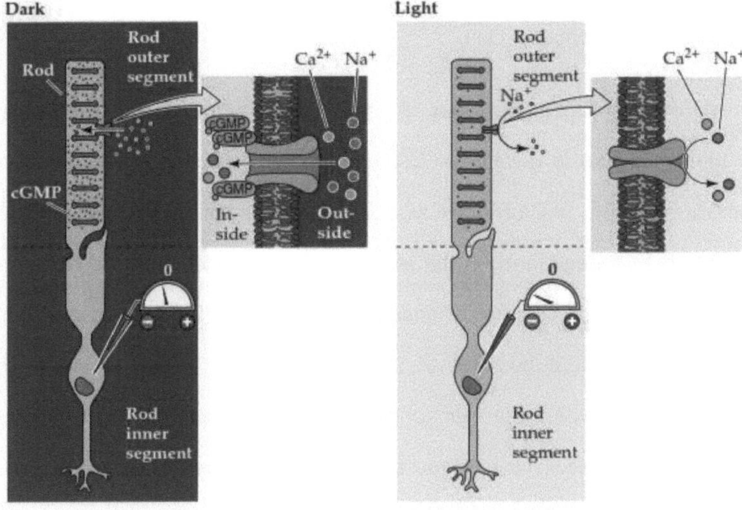

Figure 16: In the dark, cGMP levels in the outer segment are high; this molecule binds to the Na⁺ channels in the membrane, allowing sodium (and other cations) to enter, thus depolarizing the cell. Exposure to light leads to a decrease in cGMP levels, a closing of the channels, and receptor hyperpolarization (Purves, 2004).

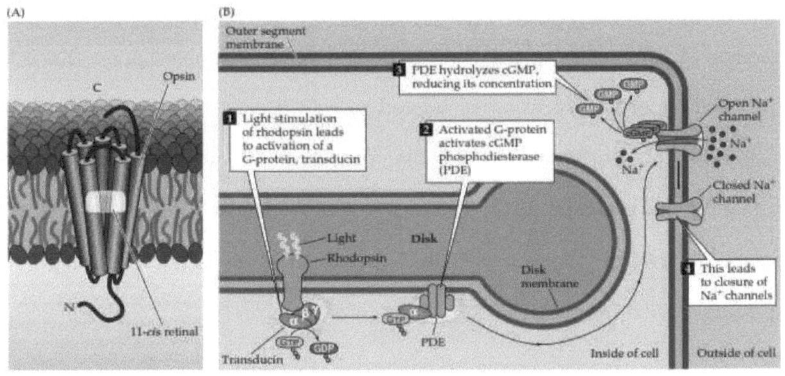

Figure 17: Details of phototransductio in rod photoreceptors. **A:** the molecular structure of rhodopsin. **B:** the phototransduction cascade (Purves, 2004).

As a result of photoreceptor hyperpolarization, less neurotransmitter (glutamate) is released into the synaptic cleft.

As previously mentioned, two types of bipolar cells can be identified depending upon their response to glutamate. ON-type bipolar cells express metabotropic glutamate receptors (mGluR) that sign invert the photoreceptor signal (Hack, 1999); i.e, they hyperpolarize in the absence of light because of the closure of mGluR-coupled cation channels (Kaneda, 2013) and depolarize when stimulated by light (Soucy, 1998) because of the inactivation of mGluR owing to the reduction of glutamate input from photoreceptors (Kaneda, 2013). Therefore, mGluR-coupled intracellular cascade becomes inactive, removing the inhibition from the cation channel and inducing cell depolarization (Kaneda, 2013).

OFF-type bipolar cells express ionotropic glutamate receptors (AMPA) (α-amino-3-hydroxy-5-methyl-4-isoxazolepropionic acid kainic acid type (KA)) that sign conserve the photoreceptor signal (Hack, 1999); i.e, they depolarize in the absence of light because the AMPA-coupled cation channels open (Kaneda, 2013) and hyperpolarize when stimulated by light (Soucy, 1998) because of the inactivation of AMPA owing to the reduction of glutamate input from photoreceptors (Kaneda, 2013). Therefore, AMPA-coupled intracellular cascade becomes inactive, inhibiting the cation channel and inducing cell hyperpolarization (Kaneda, 2013).

Although the mammalian cone photoreceptors are thought to connect to many different types of ON- and OFF-bipolar cells, rod photoreceptors connect to only one type of bipolar cells, the rod ON-bipolar cells (**Figure 18**) (Hack 1999).

In addition to these direct connections from photoreceptors, each bipolar cell receives some of its afferences from horizontal cells. These horizontal cells are in turn connected to a set of photoreceptors that surround the central group which is directly connected to bipolar cells. As a result, the receptive field of a bipolar cell has two components: a central receptive field that travels directly from the photoreceptors to the bipolar cell, and a peripheral or surround receptive field composed of information that arrives via the horizontal cells (Schiller, 2010).

Horizontal cells act to increase the acuity of sensory signals by that lateral interaction with many photoreceptors. In fact, when light reaches the retina, some photoreceptors may be illuminated brightly while others may be illuminated much less. By suppressing the signals from these less illuminated photoreceptors, the horizontal cells ensure that only the signal from the well lit photoreceptors reaches the ganglion cells, thus improving the acuity and increasing the contrast of a visual signal (Thoreson, 2008).

Because rod bipolar cells have no direct synaptic contact with ganglion cells, amacrine cells are particularly important in mediating this interaction (Kaneda, 2013). Most commonly is the AII type of amacrine cells. Each of these amacrine cells makes two major kinds of contacts: one is a gap synaptic junction with ON-cone bipolar cells and the other is a glycinergic synaptic junction with OFF-ganglion cells (Schiller, 2010). Therefore, AII amacrine cells, when excited by synaptic inputs from rod bipolar cells, send excitatory signals to the axon terminals of ON-cone bipolar cells via gap junctions in a sign-conserving manner and send inhibitory signals to the axon terminals of OFF-cone bipolar cells through glycinergic synapses in a sign-inverting manner (Kaneda, 2013).

It has been proposed that rod signals can also travel by another route that circumvents the rod bipolar cell—namely, by passing through gap junctions to cones and from there to the cone ON and OFF bipolar cells (**Figure 18**) (Soucy, 1998).

This arrangement appears to accomplish for the rods in the inner retina what is accomplished for the cones in the outer retina: it turns the single-ended photoreceptor system into a double-ended one (Schiller, 2010). Thus, amacrine cells provide an alternative, indirect path between rod bipolar cells and ganglion cells for higher processing.

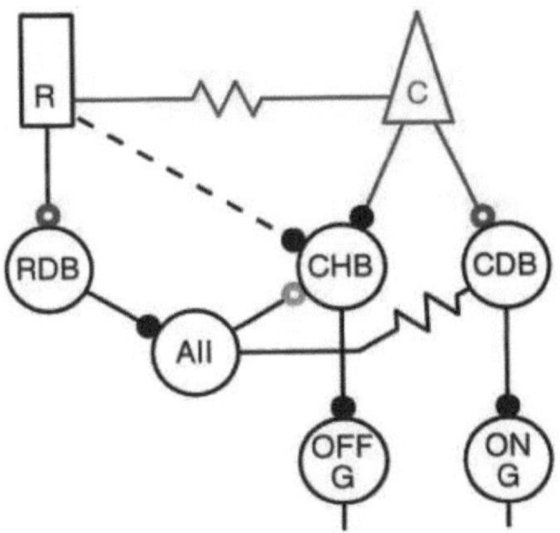

Figure 18: Schematic of the Mammalian Rod and Cone Pathways (Soucy, 1998).

Each rod bipolar cell is contacted by many rods, as many as 100 rod cells, and many rod bipolar cells contact a given amacrine cell. In contrast, the cone system is much less convergent. Thus, each retinal ganglion cell that dominates central vision (called midget ganglion cells) receives input from only one cone bipolar cell, which, in turn, is contacted by a single cone. Convergence makes the rod system a better detector of light, because small signals from many rods are pooled to generate a large response in the bipolar cell. At the same time, convergence reduces the spatial resolution of the rod system, since the source of a signal in a rod bipolar cell or retinal ganglion cell could have come from anywhere within a relatively large area of the retinal surface. The one-to-one relationship of cones to bipolar and ganglion cells is, of course, just what is required to maximize acuity (Purves, 2004).

As bipolar cells, each ganglion cell responds to stimulations of a small circular patch of the retina, which defines the cell's receptive field. The receptive fields of retinal ganglion cells (RGCs) are concentric, consisting roughly of circular central area and a surrounding ring (Purves, 2004).

RGCs have 2 basic types of receptive fields: ON-center/OFF-surround and OFF-center/ON-surround. The center and its surround are always antagonistic (Schiller, 2010).

Turning on a spot of light in the receptive field center of an ON-center ganglion cell produces a burst of action potentials. The same stimulus applied to the receptive field center of an OFF-center ganglion cell reduces the rate of discharge, and when the spot of light is turned off, the cell responds with a burst of action potentials (**Figure 19-A**). Complementary patterns of activity are found for each cell type when a dark spot is placed in the receptive field center (**Figure 19-B**). Thus, ON-center cells increase their discharge rate to luminance increments in the receptive field center, whereas OFF-center cells increase their discharge rate to luminance decrements in the receptive field center. Changes in light intensity, whether increases or decreases, are always conveyed to the brain by an increased number of action potentials.

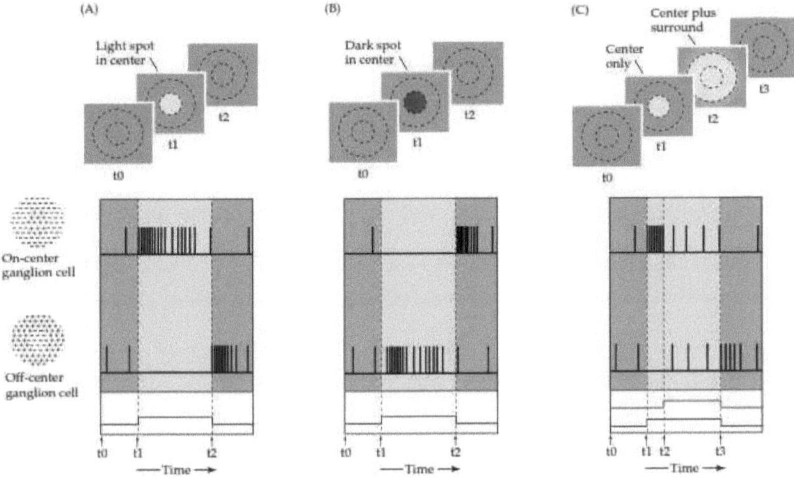

Figure 19: The responses of ON-center and OFF-center retinal ganglion cells to stimulation of different regions of their receptive fields. Upper panels indicate the time sequence of stimulus changes. (A) Effects of light spot in the receptive field center. (B) Effects of dark spot in the receptive field center. (C) Effects of light spot in the center followed by the addition of light in the surround (Purves, 2004).

The principal difference between ganglion cells and bipolar cells lies in the nature of their electrical response. In fact, intracellular recording studies revealed that photoreceptors, horizontal cells, and bipolar cells produce only graded potentials. Action potentials are first seen in some amacrine cells and are produced by all ganglion cells (Schiller, 2010).

The center of a ganglion cell receptive field is surrounded by a concentric region that, when stimulated, antagonizes the response to stimulation of the receptive field center (**Figure 19-C**). For example, as a spot of light is moved from the center of the receptive field of an ON-center cell toward its periphery, the response of the cell to the spot of light decreases (**Figure 20**). When the spot falls completely outside the center (that is, in the surround), the response of the cell falls below its resting level; the cell is effectively inhibited until the distance from the center is so great that the spot no longer falls on the receptive field at all, in which case the cell returns to its resting level of firing (Purves, 2004).

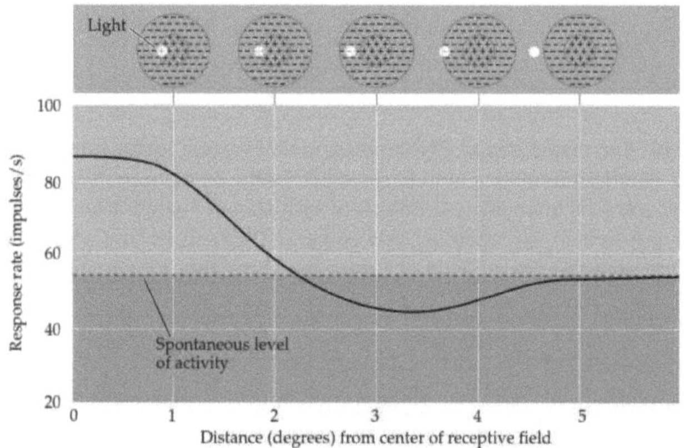

Figure 20: Rate of discharge of an on-center ganglion cell to a spot of light as a function of the distance of the spot from the receptive field center. Zero on the *x* axis corresponds to the center; at a distance of 5°, the spot falls outside the receptive field (Purves, 2004).

Like the mechanism responsible for generating the ON- and OFF-center response, the antagonistic surround of ganglion cells is a product of interactions that occur at the early stages of retinal processing (**Figure 21**) (Purves, 2004).

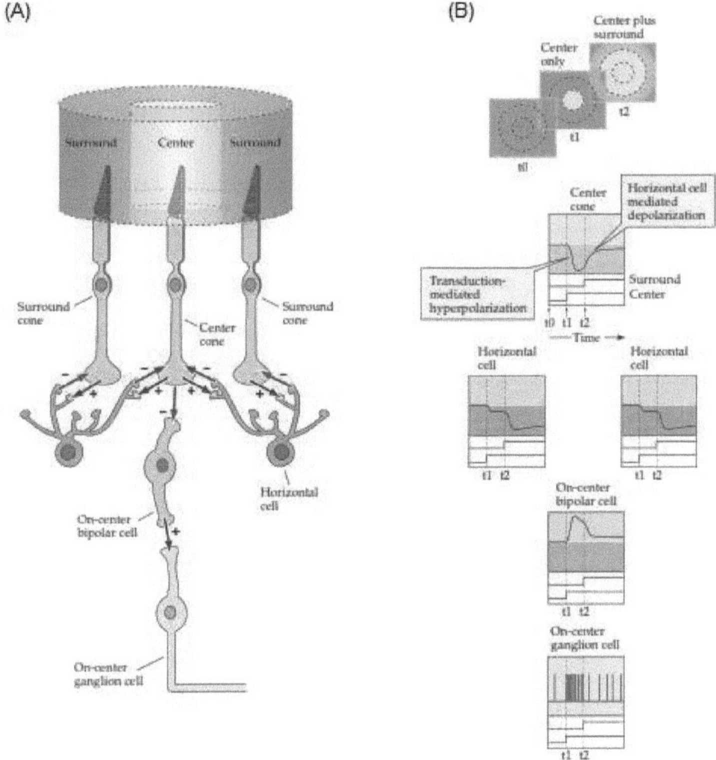

Figure 19: Circuitry responsible for generating the receptive field surround of an ON-center retinal ganglion cell. (A) Functional anatomy of horizontal cell inputs responsible for surround antagonism. A plus indicates a sign-conserving synapse; a minus represents a sign-inverting synapse. (B) Responses of various cell types to the presentation of a light spot in the center of the receptive field (t1) followed by the addition of light stimulation in the surround (t2). Light stimulation of the surround leads to hyperpolarization of the horizontal cells and a decrease in the release of inhibitory transmitter (GABA) onto the photoreceptor terminals. The net effect is to depolarize the center cone terminal, offsetting much of the hyperpolarization induced by the transduction cascade in the center cone's outer segment (Purves, 2004).

IV.3.b. Visual Path toward the Brain

The axons of the ganglion cells form the output from the retina, where they bundle together to form the **optic nerve** (which is cranial nerve II). This region of the retina contains no photoreceptors and, because it is insensitive to light, produces the perceptual phenomenon known as the blind spot (Purves, 2004). The optic disk is easily identified as a whitish circular area when the retina is examined with an ophthalmoscope; it is also recognized as the site from which the ophthalmic artery and veins enter (or leave) the eye (**Figure 20**).

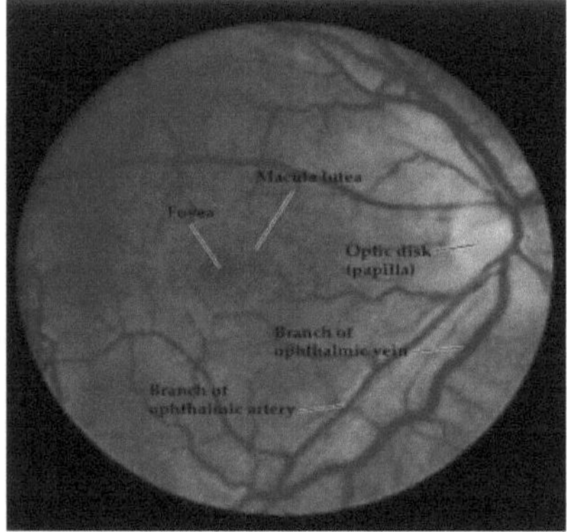

Figure 20: The retinal surface of the left eye, viewed with an ophthalmoscope. The optic disk is the region where the ganglion cell axons leave the retina to form the optic nerve; it is also characterized by the entrance and exit, respectively, of the ophthalmic arteries and veins that supply the retina (Purves, 2004).

Within the nerves, different axons carry different parts of the visual signal. Some axons constitute the magnocellular (big cell) pathway, which carries information about form, movement, depth, and differences in brightness. Other axons constitute the parvocellular (small cell) pathway, which carries information on color and fine detail. Some visual information projects directly back into the brain, while other information crosses to the opposite side of the brain. This crossing of optical pathways produces the distinctive **optic chiasma** (Greek, for "crossing") found at the base of the diencephalon, just anteriorly to the stalk of the pituitary gland (Breedlove, 2007). In humans, about 60% of these fibers cross in the chiasm, while the other 40% continue toward the thalamus and midbrain targets on the same side (Purves, 2004). Axons from the nasal retina cross over to the opposite side of the brain. Axons from the temporal retina project to their own side of the head (Breedlove, 2007). Once past the chiasm, the ganglion cell axons on each side form the **optic tract**. Thus, the optic tract, unlike the optic nerve, contains fibers from both eyes. The partial crossing (or decussation) of ganglion cell axons at the optic chiasm allows information from corresponding points on the two retinas to be processed by approximately the same cortical site in each hemisphere (Purves, 2004).

Optic nerve fibers project to several structures in the brain (**Figure 21**), the largest numbers of which pass to the thalamus (Widmaier, 2008), the routing station for all incoming sensory impulses except smell (Purves, 2004), and specifically to the lateral geniculate nuceus (LGN) (Widmaier, 2008).

Neurons in the lateral geniculate nucleus, like their counterparts in the thalamic relays of other sensory systems, send their axons to the cerebral cortex via the internal capsule. These axons pass through a portion of the internal capsule called the **optic radiation** (**Figure 21**) and terminate in the primary visual cortex (also referred to as striate cortex, Brodmann's area 17, or V1), which lies largely along and within the calcarine fissure in the occipital lobe (Purves, 2004). Different aspects of visual information are carried in parallel pathways and are processed simultaneously in a number of independent ways in different parts of the cerebral cortex before they are reintegrated to produce the conscious sensation of sight and the perception associated with it (Widmaier, 2008).

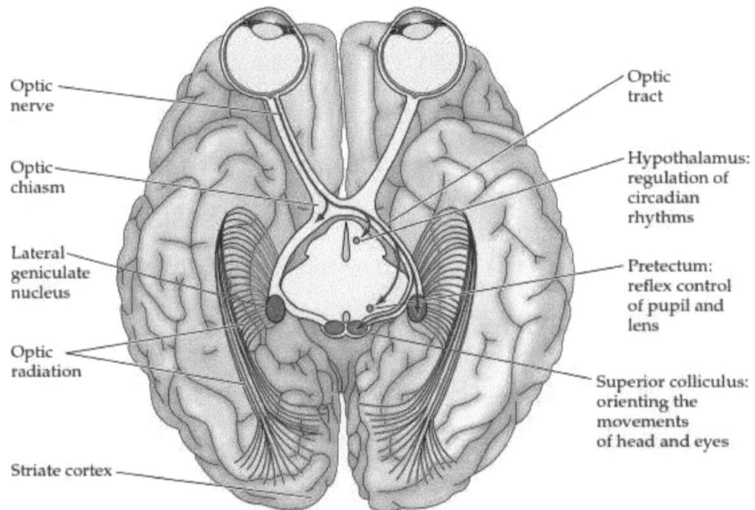

Figure 21: Central projections of retinal ganglion cells. Ganglion cell axons terminate in the lateral geniculate nucleus of the thalamus, the superior colliculus, the pretectum, and the hypothalamus (Purves, 2004).

In addition to input from the retina, many neurons of the lateral geniculate nucleus also receive input from the brainstem reticular formation and input relayed back from the visual cortex. These nonretinal inputs can control the transmission of information from the retina to the visual cortex and may be involved in our ability to shift attention between vision and other sensory modalities (Widmaier, 2008).

A second major target of the ganglion cell axons is a collection of neurons that lies between the thalamus and the midbrain in a region known as the pretectum. Although small in size compared to the lateral geniculate nucleus, the pretectum is particularly important as the coordinating center for the pupillary light reflex (i.e., the reduction in the diameter of the pupil that occurs when sufficient light falls on the retina) (**Figure 22**). The initial component of the pupillary light reflex pathway is a bilateral projection from the retina to the pretectum. Pretectal neurons, in turn, project to the Edinger Westphal nucleus, a small group of nerve cells that lies close to the nucleus of the oculomotor nerve (cranial nerve III) in the midbrain. The Edinger-Westphal nucleus contains the preganglionic parasympathetic neurons that send their axons via the oculomotor nerve to terminate on neurons in the ciliary ganglion. Neurons in the ciliary ganglion innervate the constrictor muscle in the iris, which decreases the diameter of the pupil when activated. Shining light in the eye thus leads to an increase in the activity of pretectal neurons, which stimulates the Edinger-Westphal neurons and the ciliary ganglion neurons they innervate, thus constricting the pupil (Purves, 2004).

There are several other important targets of retinal ganglion cell axons. One is the suprachiasmatic nucleus (SCN) of the hypothalamus, a small group of neurons at the base of the diencephalon. The SCN is a cluster of cells that is considered to be the body's internal clock, which controls our circadian cycle. The SCN sends information to the pineal gland, which is important in sleep/wake patterns and annual cycles (Purves, 2004).

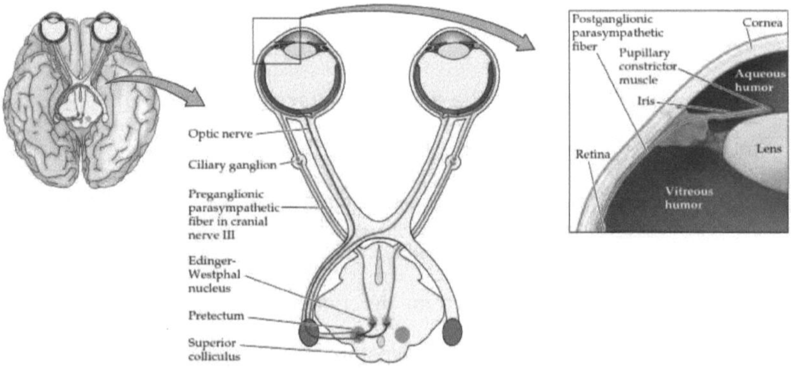

Figure 22: The circuitry responsible for the pupillary light reflex (Purves, 2004)

Another target is the superior colliculus, a prominent structure visible on the dorsal surface of the midbrain (**Figure 23**). The superior colliculus coordinates head and eye movements to visual (as well as other) targets. It is the site where eye movements are coordinated and integrated with auditory information.

IV.3.c. Visual field

The part of the world that you can see without moving your head or eyes is called your **visual field**. Each eye sees only a portion of the visual field (**Figure 23**) (Breedlove, 2007).

The retina can be vertically divided into nasal (next to you nose) and temporal (next to your temple) divisions and horizontally into superior and inferior divisions. Corresponding vertical and horizontal lines in visual space intersect at the point of fixation (the point in visual space that falls on the fovea) and define the quadrants of the visual field (Purves, 2004).

The visual field can be divided into right and left visual hemifields. The right visual hemifield is seen by the temporal left retina and nasal right retina, while the left visual hemifield is seen by the nasal left retina and the temporal right retina. The visual fields of both eyes overlap extensively in the central portion of each visual hemifield. This region defines the binocular field of view. Vision in the periphery of the field of view is strictly monocular, mediated by the most medial portion of the nasal retina (Breedlove, 2007).

The crossing of light rays diverging from different points on an object at the pupil causes the images of objects in the visual field to be inverted and left-right reversed on the retinal surface. As a result, objects in the temporal part of the visual field are seen by the nasal part of the retina, and objects in the superior part of the visual field are seen by the inferior part of the retina (Purves, 2004).

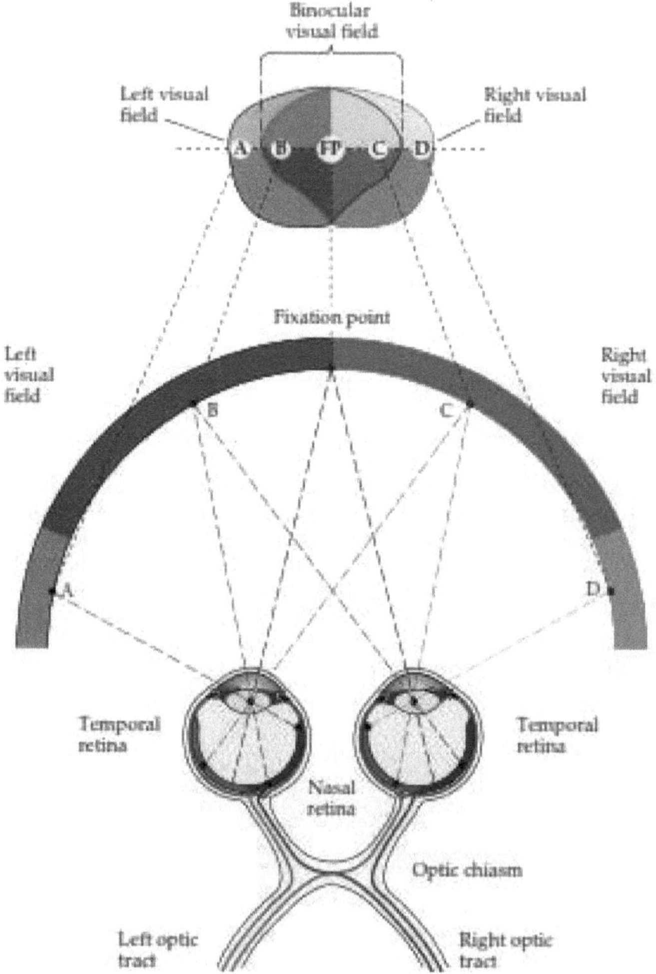

Figure 23: Projection of the binocular field of view onto the two retinas and its relation to the crossing of fibers in the optic chiasm. Information from the left visual field is carried in the right optic tract, and information from the right visual field is carried in the left optic tract (Purves, 2004).

IV.3.d. Light Adaptation versus Dark Adaptation

If you move from a place of bright sunlight into a darkened room, a temporary "blindness" takes place until the photoreceptors can undergo **dark adaptation**. In the low levels of illumination of the darkened room, vision can be only supplied by the rods, which have greater sensitivity than the cones. During the exposure to bright light, however, the rods' rhodopsin has been completely activated, making the rods insensitive to light (Widmaier, 2008). Rhodopsin cannot respond fully again until it is restored to its resting state, a process requiring several minutes in which rhodopsin is turned off in two steps, first it is phosphorylated by rhodopsin kinase, and this then allows the binding of a quenching protein called arrestin (Lagendo, 2014). Dark adaptation occurs, in part, as enzymes regenerate the initial form of rhodopsin, which can respond to light. Vitamin A is necessary for good night vision because it is required for the synthesis of the retinal portion of the rhodopsin (Widmaier, 2008).

In contrast to dark adaptation, **light adaptation** occurs when you step from a dark place into a bright one. Initially, the eye is extremely sensitive to light, and the visual image is too bright and has poor contrast. However, the rhodopsin is soon used up, or "bleached" by the bright light, and the rods become unresponsive so that only the less sensitive cones are operating and the image becomes less bright (Widmaier, 2008).

V. Most Common Eye Disorders and Diseases

Diseases of the eye are so common. Millions of people have visual problems. Although minor visual problems are not even considered to be true diseases, many of the eye diseases, if left untreated, can cause blindness. Here is a brief description of some common eye diseases.

V.1. Disorders of Alignment and Movement

Normal vision requires that both eyes be precisely aligned and move together in strict coordination. **Strabismus** is a disorder in which the two eyes do not line up in the same direction, and therefore do not look at the same object at the same time. The condition is more commonly known as "crossed eyes" (McConnell, 2007).

Six different muscles surround each eye and work as a team, so that both eyes can focus on the same object. In people with strabismus, these muscles do not work together. As a result, one eye looks at one object, while the other eye turns in a different direction and is focused on another object (McConnell, 2007).

When this occurs, two different images are sent to the brain, one from each eye. This confuses the brain. In children, the brain may learn to ignore the image from the weaker eye (McConnell, 2007).

If the strabismus is not treated, the eye that the brain ignores will never see well. This loss of vision is called **amblyopia**. Another name for amblyopia is "lazy eye" (McConnell, 2007).

Nystagmus is rapid involuntary, repetitive motion of one or both eyes. It may be caused by neurological disease (especially of the cerebellum), disease of the inner ear, or drug toxicity (McConnell, 2007).

V.2. Disorders of Refraction

Refractive disorders are those in which images are blurred by failure of light rays to focus precisely on the retina. These abnormalities may be caused by abnormal shape of the cornea, elongation or shortening of the eyeball, or stiffness of the lens which interferes with focus adjustment. These disorders are not usually associated with pathological abnormalities.

In some cases, the image of distant objects focuses in front of, rather than on, the retina, a condition called **myopia (Figure 24)** (McConnell, 2007). It can be caused by the corneal surface being too curved, or by the eyeball being too long (Purves, 2004).

Myopia is commonly referred to as shortsightedness, is a common cause of visual disability throughout the world. It is so called because the lens is usually able to overcome the defect for near objects (Fredrick, 2002).

People with myopia can be classified in two groups, those with low to modest degrees of myopia (referred to as "simple" or "school" myopia, 0 to - 6 dioptres*) and those with high or pathological myopia (greater than - 6 dioptres). Simple myopia can be corrected with spectacles or contact lenses, whereas "high" (pathological) myopia is often associated with potentially blinding conditions such as retinal detachment, macular degeneration, and glaucoma (Fredrick, 2002).

In foreshortened vision, the retina is too close to the cornea, the focal point is behind the retina, and the image is blurred, a condition called **hyperopia**. People with hyperopia are said to be farsighted because the lens is often able to overcome the defect for distant objects **(Figure 24)** (McConnell, 2007).

For focus on distant objects, ciliary muscles pull at the edges of the lens and flatten it; for near objects, the muscles relax and allow the lens to assume a thicker, more globular shape. Focusing power is gradually lost over the years and results in a slow decrease in the ability of the eye to focus on nearby objects, a condition called **presbyopia**. About age of 40, people gradually notice an inability to focus on near objects as they find the need to hold reading materials further away. It is a natural part of the aging process and affects everyone (McConnell, 2007).

Sometimes, the curvature of the cornea is not uniform. In these cases, the image is properly focused in some areas and blurred in others, a condition called **astigmatism (Figure 24)** (McConnell, 2007).

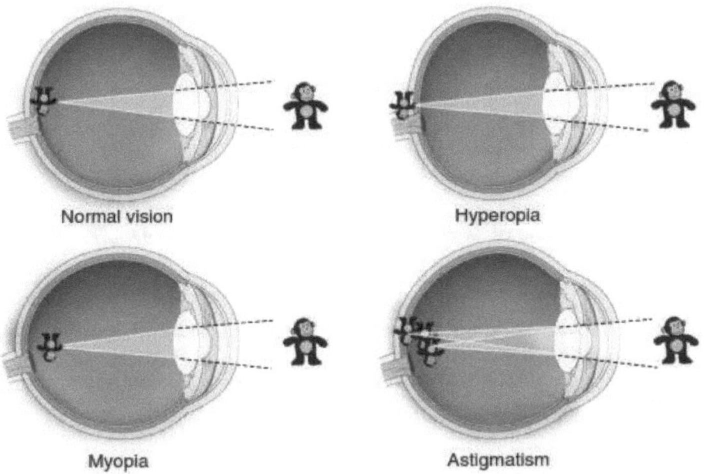

Figure 24: Refraction of the Eye. Normal vision versus three common irregularities (Leonard, 2014).

These refractive disorders are corrected by eyeglasses, contact lenses, or surgical reshaping of the cornea.

V.3. Disorders of the Orbit

The forward displacement of the globe in the orbit, or bulging eyes, is **proptosis**. This condition is caused by increased tissue and fluid in the orbit behind the globe. An important cause of proptosis is Graves disease (hyperthyroidism) that causes accumulation of fat and fluid behind the globe and produces a remarkable "bug-eyed" appearance. Protosis is a clinical finding that also occurs with orbital inflammation, infection, or mass, such as hemangioma, lymphoma, or lacrimal gland adenoma (McConnell, 2007). Additional irregularities or abnormal conditions of the eye include the followings:

V.4. Cataract

Cataract is an abnormal progressive condition of the lens, characterized by loss of transparency; it is clouding of the lens. Most cataracts are caused by age-related degeneration of lens fibers, in which the fibers break into molecules small enough to exert an osmotic effect that attracts water into the lens, clouding it and interfering with light transmission. Cataracts are a major cause of poor vision and blindness around the world (McConnell, 2007).

Diabetes can be a cause of cataract, in which excessive blood sugar is converted in the lens into sorbitol, a molecule with high osmotic power that attracts water, thereby clouding the lens (McConnell, 2007).

V.5. Retinitis Pigmentosa (RP)

Retinitis pigmentosa refers to a heterogeneous group of hereditary eye disorders characterized by progressive vision loss due to a gradual degeneration of photoreceptors (purves, 2004).

V.6. Age-Related Macular Degeneration (ARMD)

Macular degeneration is an eye disorder that slowly destroys sharp, central vision. This makes it difficult to see fine details and read. This disease is most common in people over age 60, which is why it is often called age-related macular degeneration (ARMD) (Purves, 2004).

V.7. Diabetic Retinopathy

Diabetic retinopathy is retinopathy (damage to the retina) caused by complications of diabetes, which can eventually lead to blindness. In this condition, retinal arteries develop microaneurysms and leak plasma or blood, creating retinal exudates and hemorrhages (McConnell, 2007).

V.8. Glaucoma

Glaucoma is not a singular eye disease, but is instead a term for several eye conditions that can damage the optic nerve. Glaucoma is usually the result of abnormally high pressure inside the eye. Over time, the increased pressure can erode the optic nerve tissue, which may lead to vision loss or even blindness. If caught early, additional vision loss may be prevented.

V.9. Diet and Vision

Certain nutrients in our diet play a significant role in maintaining healthy vision by preventing many eye diseases and disorders.

Some of these eye-essential nutrients include vitamins C and E, carotenoids, omega-3 fatty acids and the trace mineral zinc. Eating a well-balanced diet based on whole foods should ensure that you're meeting your daily needs for these nutrients that help maintain your eyesight (Rasmussen & Johnson, 2013).

Getting enough vitamin C and vitamin E may keep your eyesight at its best. These vitamins function as antioxidants, which help to protect your eyes from damage caused by harmful free radicals. Over time, this may prevent or slow the progression of cataracts and macular degeneration. You can increase your intake of vitamin E by consuming more wheat germ, nuts, sunflower seeds, almonds, hazelnuts and peanuts, and spinach. You'll get vitamin C from bell peppers, citrus fruits, broccoli, strawberries and tomatoes (Christen, 2008).

Compounds called lutein and zeaxanthin have a positive impact on your sight. They're a class of plant chemicals called carotenoids found in peas, broccoli, green beans, eggs and leafy greens such as spinach. The retina of your eye contains a high concentration of both lutein and zeaxanthin, which protect your eyesight by acting as antioxidants and by filtering harmful wavelengths of light (Koushan, 2013).

Eicosapentaenoic acid, or EPA, and docosahexaenoic acid, or DHA, which are types of omega 3 fatty acids (ω3FAs), may prevent retinopathies. Abnormal blood vessel growth in the retina of your eye causes retinopathies and may cause blindness. Getting enough EPA and DHA from cold-water fish, such as salmon, tuna and mackerel may halt this abnormal growth and keep your eyesight sharp (Egilmez, 2008).

A lack of the mineral zinc in your body may also cause declining eyesight. Zinc is highly concentrated in the retina of your eye, and it plays a role in production of melanin, a pigment that protects your eyes. A deficiency of zinc may cause poor night vision, cloudy cataracts and macular degeneration. Zinc is important in maintaining the health of the retina, given that zinc is an essential constituent of many enzymes and needed for optimal metabolism of the eye. Zinc ions are present in the enzyme superoxide dismutase, which plays an important role in scavenging superoxide radicals. As related to the eye, zinc plays important roles in antioxidant and immune function. Zinc also plays an important role in the structure of proteins and cell membranes. The structure and function of cell membranes are also affected by zinc. Loss of zinc from biological membranes increases their susceptibility to oxidative damage and impairs their function. Zinc also plays a role in cell signaling and has been found to influence nerve-impulse transmission. You can get more zinc by eating red meat, nuts, seafood, eggs, beans, yogurt and poultry (Rasmussen & Johnson, 2013).

Refrences

1. **Breedlove, S., Rosenzweig, M., & Watson, N.** (2007). *Biological Psychology: An Introduction to Behavioral, Cognitive, and Clinical Neuroscience, 5th edition.*

2. **Cakiner-Egilmez, T.** (2007). Omega 3 fatty acids and the eye. *Insight (American Society of Ophthalmic Registered Nurses), 33*(4), 20-5.

3. **Christen, W. G., Liu, S., Glynn, R. J., Gaziano, J. M., & Buring, J. E.** (2008). A prospective study of dietary carotenoids, vitamins C and E, and risk of cataract in women. *Archives of ophthalmology, 126*(1), 102.

4. **Demb, J. B., & Singer, J. H.** (2012). Intrinsic properties and functional circuitry of the AII amacrine cell. *Visual neuroscience, 29*(01), 51-60.

5. **Fredrick, D. R.** (2002). Myopia. *BMJ: British Medical Journal, 324*(7347), 1195.

6. **Hack, I., Peichl, L., & Brandstätter, J. H.** (1999). An alternative pathway for rod signals in the rodent retina: rod photoreceptors, cone bipolar cells, and the localization of glutamate receptors. *Proceedings of the National Academy of Sciences, 96*(24), 14130-14135.

7. **Hildebrand, G. D., & Fielder, A. R.** (2011). Anatomy and Physiology of the Retina. In *Pediatric retina* (pp. 39-65). Springer Berlin Heidelberg.

8. **Isayama, T., Zimmerman, A., & Makino, C.,** (2013). *The molecular design of visual transduction.*

9. **Kaneda, M.** (2013). Signal processing in the Mammalian retina. *Journal of Nippon Medical School, 80*(1).

10. **Kefalov, V. J.** (2012). Rod and cone visual pigments and phototransduction through pharmacological, genetic, and physiological approaches. *Journal of Biological Chemistry, 287*(3), 1635-1641.

11. **Kolb, H., Fernandez, E., & Nelson, R.,** (2013). *Webvision: the organization of the retina and visual system.*

12. **Koushan, K., Rusovici, R., Li, W., Ferguson, L. R., & Chalam, K. V.** (2013). The role of lutein in eye-related disease. *Nutrients, 5*(5), 1823-1839.

13. **Lamb, T. D., & Pugh, E. N.** (2006). Phototransduction, dark adaptation, and rhodopsin regeneration the proctor lecture. *Investigative ophthalmology & visual science, 47*(12), 5138-5152.

14. **Leonard, P.,** (2014). *Quick and Easy Medical Terminology, 7th edition.*

15. **Luo, D. G., Xue, T., & Yau, K. W.** (2008). How vision begins: an odyssey. *Proceedings of the National Academy of Sciences, 105*(29), 9855-9862.

16. **Luo, D. G., Yue, W. W., Ala-Laurila, P., & Yau, K. W.** (2011). Activation of visual pigments by light and heat. *Science, 332*(6035), 1307-1312.

17. **Masland, R. H.** (2012). The neuronal organization of the retina. *Neuron, 76*(2), 266-280.

18. **McConnell, T.,** (2007). *The Nature Of Disease: Pathology For The Health Professions, 1st edition*

19. **Menon, S. T., Han, M., & Sakmar, T. P.** (2001). Rhodopsin: structural basis of molecular physiology. *Physiological reviews, 81*(4), 1659-1688.

20. **Purves, D., Augustine, G., Fitzpatrick, D., Hall, W., Anthony, S., Lamantia, A., Mcnamara, J., & Williams, S.,** (2004). *Neuroscience, 3rd edition.*

21. **Rasmussen, H. M., & Johnson, E. J.** (2013). Nutrients for the aging eye. *Clinical interventions in aging, 8,* 741.

22. **Schiller, P. H.** (2010). Parallel information processing channels created in the retina. *Proceedings of the National Academy of Sciences, 107*(40), 17087-17094.

23. **Schwartz, E. A.** (2002). Transport-mediated synapses in the retina. *Physiological reviews, 82*(4), 875-891.

24. **Shichida, Y.** (2009). Evolution of opsins and phototransduction. *Philosophical Transactions of the Royal Society B: Biological Sciences, 364*(1531), 2881-2895.

25. **Shichida, Y., & Yamashita, T.** (2003). Diversity of visual pigments from the viewpoint of G protein activation—comparison with other G protein-coupled receptors. *Photochemical & photobiological sciences, 2*(12), 1237-1246.

26. **Solomon, S. G., & Lennie, P.** (2007). The machinery of colour vision. *Nature Reviews Neuroscience, 8*(4), 276-286.

27. **Soucy, E., Wang, Y., Nirenberg, S., Nathans, J., & Meister, M.** (1998). A novel signaling pathway from rod photoreceptors to ganglion cells in mammalian retina. *Neuron, 21*(3), 481-493.

28. **Sterling, P.** (2013). Some Principles of Retinal Design: The Proctor Lecture. *Investigative ophthalmology & visual science, 54*(3), 2267-2275.

29. **Sung, C. H., & Chuang, J. Z.** (2010). The cell biology of vision. *The Journal of cell biology, 190*(6), 953-963.

30. **Thoreson, W. B., Babai, N., & Bartoletti, T. M.** (2008). Feedback from horizontal cells to rod photoreceptors in vertebrate retina. *The Journal of Neuroscience, 28*(22), 5691-5695.

31. **Widmaier, E., Raff, H., Strang, K.,** (2008). *Vander's Human Physiology: the Mechanism of Body Function, 11th edition.*

32. www.allaboutvision.com/resources/anatomy.htm. Retrieved on 10/4/2018

33. http://lumenistics.com/what-is-full-spectrum-lighting/. Retrieved on 6/5/2018

34. http://www.about-vision.com/refractive-errors/diopters. Retrieved on 10/5/2018

I want morebooks!

Buy your books fast and straightforward online - at one of world's fastest growing online book stores! Environmentally sound due to Print-on-Demand technologies.

Buy your books online at
www.morebooks.shop

Kaufen Sie Ihre Bücher schnell und unkompliziert online – auf einer der am schnellsten wachsenden Buchhandelsplattformen weltweit! Dank Print-On-Demand umwelt- und ressourcenschonend produzi ert.

Bücher schneller online kaufen
www.morebooks.shop

KS OmniScriptum Publishing
Brivibas gatve 197
LV-1039 Riga, Latvia
Telefax: +371 686 204 55

info@omniscriptum.com
www.omniscriptum.com

Printed by Books on Demand GmbH, Norderstedt / Germany